A Farmers' Jury

A Farmers' Jury

The Future of Smallholder Agriculture in Zimbabwe

Stuart A. Coupe, Jon Hellin,
Absolom Masendeke, Elijah Rusike

Practical Action Publishing Ltd
27a Albert Street, Rugby, CV21 2SG, Warwickshire, UK
www.practicalactionpublishing.org

© Intermediate Technology Publications 2005

First published 2005

ISBN 10: 1 85339 576 5
ISBN 13: 9781853395765
ISBN Library Ebook: 9781780441474
Book DOI: http://dx.doi.org/10.3362/9781780441474

A catalogue record for this book is available from the British Library.

The authors, contributors and/or editors have asserted their rights under the Copyright
Designs and Patents Act 1988 to be identified as authors of their respective contributions.

Since 1974, Practical Action Publishing has published and disseminated books and
information in support of international development work throughout the world. Practical
Action Publishing is a trading name of Practical Action Publishing Ltd (Company Reg.
No. 1159018), the wholly owned publishing company of Practical Action. Practical Action
Publishing trades only in support of its parent charity objectives and any profits are
covenanted back to Practical Action (Charity Reg. No. 247257, Group VAT Registration No.
880 9924 76).

Typeset by J&L Composition, Filey, North Yorkshire

Contents

Acknowledgements

This publication is a result of a collaborative effort involving Intermediate Technology Development Group Southern Africa Regional and UK based international staff. The actual jury process was facilitated with great skill and professionalism by Elijah Rusike. He was strongly supported by Stanley Chisimbe and Julius Mugwagwa from the Biotechnology Trust of Zimbabwe.

It is particularly important to thank all Senior Government Officials, academics, and other experts and representatives of farmers' organizations who demonstrated unparalleled commitment to the jury deliberations and who worked so hard to make it a success. Without their participation and contributions, this event would have been a total flop. There were many anxious moments during the process when exchanges tended to be heated, but all the participants remained focused on the issues and visions for the future. Such unity of purpose deserves to be applauded.

Members of the Oversight Panel, in particular Dr Unesu Ushewokunze, showed commitment, impartiality, and fairness towards the whole process despite heavy work schedules.

Special thanks go to the great moral, technical, and financial support from IDS based in the UK as well as guidance and support from Tom Wakeford who has accumulated substantial experience on farmers' juries in India.

Many thanks go to Johanna Mutsambiwa for excellent quality transcriptions of many hours of video coverage from Shona into English.

The whole jury process resulted in a useful international learning network on farmers' juries that gives a great deal of hope in terms of the continued search for ideas which will unlock smallholder participation in policy formulation. It is our hope that this publication will enrich the content and direction of other participatory policy formulation initiatives across the world.

This publication is funded by the European Commission under Strengthening Livelihoods of Poor Rural Communities in Southern Africa, a project implemented by ITDG under B7-6000: Co-Financing NGO operations.

Executive summary

The Farmers' Jury event described in this paper was held in February 2003, to give a grassroots farmers' perspective on the Zimbabwean government's vision for the future of agriculture. This happened at a time when Zimbabwe was still grappling with socio-economic and political challenges emanating from the 2002 elections, fast track land reform process, food scarcity, and political polarization at all levels of Zimbabwean society. At the same time, the country was experiencing fundamental reforms in the judiciary and media sectors. As such national, regional, and international partners in the jury project were very sceptical of the possibility of Zimbabwe hosting a successful and peaceful jury process in such a context. With full knowledge of the situation in Zimbabwe, the organizers planned the jury process very carefully through strategic consultations and participation of actors in government, NGOs and the private sector. In the event it passed off peacefully, without being brought to the attention of either the national or international media.

The strategic planning and consultation process that was used to design the jury process in Zimbabwe brought a great deal of excitement to both farmers and government officials on the possibility of creating dialogue and interaction. High-ranking government officials were very receptive to the invitation to participate in the process as they felt they needed to clarify a number of issues with farmers. Before the actual jury itself, farmers held consultative meetings with their colleagues to identify issues of prime concern to them. These issues were then prioritized and experts on the issues were identified.

The dialogue between farmers and high-ranking government officials is the basis for the paper. The farmers were not intimidated by the occasion, often expressing their criticism of government policies with irony and sarcasm. This did not spark controversy at the time because it was quite clear to all those present that the farmers were giving their own perspectives. Unlike the views of NGOs, those of farmers are difficult to discredit, given that they are held up as the backbone of the Zimbabwean nation, and are the supposed beneficiaries of all key government policies.

Over five days, the farmers formed a jury to hear evidence from a range of policy-makers on the future of agriculture. In all, they cross-examined witnesses for 11 different priority topics and issued a reasoned verdict on 13 issues. This paper does not attempt to look at each topic in detail. Instead examples are given of three types of topics.

- Issues at the top of the agenda in terms of local and global policy: land reform and HIV/AIDS.
- An issue that generated the most passionate and heated exchanges of the week: the role of the Zimbabwe Farmers' Union in representing the interests of small-scale farmers.
- Topics that were quite new to most of the farmers, where they considerably advanced their understanding of the policy issues at stake in the process of reaching a verdict.

The working paper is written to appeal both to those with an interest in participatory methods and to those with an interest in the situation of small-scale peasant farmers in Zimbabwe. It is intended as a stimulus to further discussion, debate, and above all, new actions and initiatives to give these farmers a greater say in the future.

1

Small-scale farming in Zimbabwe

A vibrant agricultural sector is essential for the economic growth of developing countries. Smallholder farming is linked to reductions in rural poverty and inequality. According to the World Bank growth in agricultural incomes is particularly effective at reducing rural poverty, because it has knock-on or multiplier effects on local markets for other goods and services provided by non-farm rural poor, such as construction, manufacturing, and repairs (World Bank, 2001: 67). Strong agricultural growth has been a feature of countries, such as Bangladesh, Indonesia, and China that have successfully reduced poverty.

The agriculture sector accounts for 24 per cent of Africa's GDP, 40 per cent of its foreign exchange earnings and 70 per cent of its employment. In Zimbabwe more than 50 per cent of the population live in rural areas and are dependent on smallholder agriculture for their livelihood. Like farmers worldwide, Zimbabwean smallholder farmers face some daunting obstacles and challenges. Factors of major concern include:

- land, soil, and water resource degradation; stagnant or declining agricultural productivity, aggravated by inequitable land distribution;
- depressed international crop prices and unfair competition in domestic markets from imported products, due to subsidized agricultural over-production in the rich world;
- physical and commercial isolation from the markets and potential channels of economic growth emerging in domestic or international trade;
- inadequate access to the knowledge, technology, and skills needed to diversify rural livelihoods, secure markets for increased agricultural productivity, and manage resources on a sustainable basis.

Whilst not facing systematic discrimination in terms of access to basic services such as education, small-scale farmers in communal areas are nevertheless a disadvantaged and marginalized group in the Zimbabwean context. They live primarily on agriculture in communal lands where there is no individual land entitlement. Such areas are normally in the marginal natural regions 3, 4 and 5. These areas are characterized by low rainfall and poor granitic soils. The population density in such areas can be as high as 30 people per square kilometre. That pressure has in some cases resulted in high soil erosion and degraded environments. Infrastructure and communication is often very poor.

With Independence in 1980, Zimbabwe inherited a highly unequal distribution of land, with the majority black farming population being confined to the former 'reserves', most of which were located in the marginal agro-ecological zones. These areas were renamed 'communal lands' in 1982, but beyond a shift in nomenclature nothing much changed. In the early 1980s the new government encouraged various attempts at resettlement, but the extent of these were constrained by constitutional limitations (under the Lancaster House agreement), bureaucratic delay, and funding shortages. By the end of the 1980s the government had resettled some 52 000 households and purchased 2.7 million hectares (around 16 per cent of commercial farmland). Settlers were moved to a range of different 'scheme' types under a series of technocratic models. These were seen as separate from the communal areas with resettlement farmers expected to develop independent full-time farming operations, often far from their original homes. Despite these movements, the former reserves remained crowded with poor agricultural potential, and livelihoods continued to rely on mostly dry land farming, keeping livestock, and, significantly, remittance incomes from circular migration to towns, farms, and mines (Chaumba, Scoones, and Wolmer, 2003).

The communal areas were hit by a wave of negative external factors, notably the major droughts of the early 1980s and 1990s, which devastated cattle populations, and the changing fortunes of the economy, particularly following the implementation of the structural adjustment programme from 1991. This saw a major downturn in economic fortunes for many, especially those who relied upon remittance income from relatives living in town. Combined with these factors, HIV/AIDS has had a major impact on the demography and livelihoods of communal area populations, particularly from the mid-1990s. This has changed household structures, reducing key sections of the farming labour force, and resulting in an increasing number of female or child headed households. By the late 1990s a lack of employment opportunity, a growing ill-health burden, smaller land areas (and evidence of effective landlessness), combined with a severe lack of capital and draft power, constrained and marginalized many people.

1.1 Agricultural policy in communal areas

Government agricultural policy has always been based on the ideal of transformation of communal area agriculture 'from subsistence to commercial'. This seems to imply increased market integration, increased use of external inputs, and a concentration on those farmers able to produce a marketable surplus. While this may be suitable for a proportion of smallholders, it leaves those, probably the majority, who do not have the resources to follow this route, exposed to marginalization and lack of sustainability. (Whiteside, 1998) It also fails to consider in a holistic way, the rural household livelihood, which is often made up from both farm and off-farm components.

- Most of the agricultural establishment has been trained within the 'modernization' ethos.
- Input suppliers, who naturally emphasize the use of inputs, are doing an increasing proportion of research and demonstration.
- The Zimbabwe Farmers Union (see below) is being used to represent the needs of smallholders in research and extension planning processes. However the smallholder does not have a single perspective, and those smallholders aspiring to commercialization tend to be particularly well represented in ZFU decision-making structures.

From Independence to the end of the 1980s, the focus of state support shifted to providing extension, subsidized credit, and markets for communal areas. This support resulted in a remarkable rise in smallholder maize production during the 1980s. Since then however there has been a trend to:

- limit the Grain Marketing Board (GMB) offer price for maize;
- reduce the number of GMB purchase points in rural areas;

- sell off surplus grain on the regional/world market (this created a disaster in 1992 when 'surplus' grain was sold immediately prior to the severe drought).

Fiscal constraints in the late 1980s and 1990s led to the Economic Structural Adjustment Programme (ESAP) with attempts to restrain government expenditure and end subsidies. This had a marked effect on agriculture with the privatization of marketing parastatals, with the exception of GMB, which as noted above, has nevertheless reduced its subsidized service to communal area farmers. The reduction in subsidies to agriculture and the fall in expenditure on the Ministry of Agriculture as a percentage of overall expenditure has been dramatic.

Failing crops in a dought-affected communal area (Chivi District)   *Source*: ITDG/Kuda Murwira

2

The idea of the Farmers' Jury

The event was part of an international project to test the relevance and adaptability of citizens' juries for giving marginalized groups a voice in policy making in developing countries. The Farmers' Jury was a promotion of the citizens' jury approach with a view to its adoption in a range of potential situations in Zimbabwe, where hitherto policy formulation has remained the preserve of technocrats and politicians. The participation of the supposed primary beneficiaries of these policies has been zero, or at most superficial. In the recent past, many policies developed by the technocrats to benefit smallholder-farming communities have fallen far short of their anticipated results. For example, the government recently declared all maize to be a state controlled commodity and people were not allowed to trade in it independently. However, farmers did withhold their grain from the state marketing board, as the fixed selling price was considered low. An illegal trade in maize then flourished, resulting in rocketing prices for the consumer, the very phenomenon the original edict was designed to tackle. The failure to enter into any discussions or negotiations with small-scale farmers over what constituted a fair price led to the failure of the policy.

One of the major reasons for this failure emanates from poorly informed policy formulation processes. Processes that effectively engage the primary beneficiaries have been shown to result in relevant and progressive policies that promote sustainable development. But how can the full participation of the smallholder farming communities be achieved? What framework could guide their participation?

2.1 The citizens' jury – a deliberative and inclusive process

In both North and South, attempts to overcome low public confidence in government institutions and scientific expertise have often emphasized a more deliberative and inclusive form of debate and policy-making. Advocates argue that Deliberative and Inclusive Processes (DIPs) allow multiple perspectives into debates, thereby generating better understandings of the uncertainties of policy questions. The potential of DIPs to broaden democratic control over science and technology is also important. The main feature of Deliberative and Inclusive Processes are as follows (Irwin and Wynne, 1996):

- Deliberation is defined as 'careful consideration' or 'the discussion of reasons for and against'. Deliberation is a common, if not inherent, component of all decision-making and of democratic societies.
- Inclusion is the action of involving others and an inclusive decision-making process is based on the active involvement of many social actors, and usually emphasizes the participation of previously excluded citizens.
- Social interaction is usually part of any DIP, and normally incorporates face-to-face meetings between those involved.

- There is a dependence on language through discussion and debate. This is usually in the form of verbal and visual constructions rather than written text.
- A deliberative process assumes that, at least initially, there are different positions held by the participants and that these views should be respected.
- DIPs are designed to enable participants to evaluate and re-evaluate their positions in the light of different perspectives and new evidence.
- The form of negotiation is often seen as containing value over and above the 'quality' of the decisions that emerge. Participants share a commitment to resolving problems through public reasoning and dialogue aimed at mutual understanding, even if consensus is not being sought.
- It is recognized that, while the goal is usually to reach decisions, or at least positions upon which decisions can subsequently be taken, an unhurried, reflective and reasonably open-ended discussion is required.

Two closely related techniques for deliberative decision-making, citizens' juries and planning cells, were developed in the late 1960s and early 1970s. Compared with other DIPs, citizens' juries offer a unique combination of information, time, scrutiny, deliberation, and independence. In a citizens' jury a representative panel of citizens meets for a total of 30 to 50 hours to examine carefully an issue of public significance. The jury, of between 12 and 20 members, serves as a microcosm of the public. They hear from a variety of specialist witnesses and are able to deliberate together on an issue. On the final day of their moderated hearings, the members of the jury present their recommendations to decision-makers and the public. Citizens' juries have a number of features that distinguish them from other methods of participation:

- Participants are systematically recruited, rather than just being asked to turn up via an open invitation.
- Participants are given the opportunity to scrutinize the information that they receive from witnesses.
- Participants are given time to reflect and deliberate on the questions at hand, usually assisted by a facilitator.
- Acting as 'jurors', participants are expected to develop a set of conclusions or 'visions' for the questions before them.

The Farmers' Jury in Zimbabwe is part of an initiative to adapt the methodology to developing country conditions. Very broadly, it can be said that due to resource constraints policy formulation in developing countries is more chaotic and ad hoc. As in developed countries, policy formulation is very much in the hands of a technocratic class of civil servants, but due to the constant frustrations of working under resource constraints, these high-ranking officials seem to be much more accessible and willing to reach out and engage in new and experimental processes with civil society organizations.

2.2 The Andhra Pradesh experience

Although taking place in a different social and ecological context, the process that acted as a model for those planning a citizens' jury in Zimbabwe was Prajateerpu (meaning 'people's verdict' in the local language *Telegu*), which took place in Andhra Pradesh (AP), India, in June 2001.

Strongly backed by the World Bank and DFID as a model of progressive government in India, the AP Government issued a strategy document called Vision 2020 (Pimbert and Wakeford, 2003). The Vision 2020 document proposed profound and fundamental changes in the food system, and yet there was no process to allow the approximately 50 million small-scale farmers in the state to help shape the policy. In this context a coalition of five organizations, two in the UK, three in AP, designed and facilitated a participatory process to encourage more public debate in policy choices for agriculture.

Box 1   *Prajateerpu*

The citizens' jury. A citizens' jury made up of representatives of small and marginal farmers, small traders and food processors, and consumers. To reflect the reality of rural Andhra Pradesh, most of the members were small and marginal farmers and included indigenous people (known in India as divas). Over two-thirds of the jury members were women.

Visions of the future. Jury members were presented with three different scenarios. Each one was advocated by some key opinion leaders who attempted to show the logic behind the scenario.

Vision 1: *Vision 2020*. The Andhra Pradesh Chief Minister has put this scenario forward and has been backed by a loan from the World Bank. It proposes to consolidate small farms and rapidly increase mechanization and modernization. Production-enhancing technologies such as genetic modification will be introduced in farming and food processing, reducing the number of people on the land from 70 per cent to 40 per cent by 2020.

Vision 2: An export-based cash crop model of organic production. This vision of the future is based on proposals within IFOAM and the International Trade Center (UNCTAD/WTO) for environmentally friendly farming linked to national and international markets. This vision is also increasingly driven by the demand of supermarkets in the North to have a cheap supply of organic produce and comply with new eco-labeling standards.

Vision 3: Localized food systems. A future scenario based on increased self-reliance for rural communities, low external input agriculture, the re-localization of food production, markets, and local economies, with long-distance trade in goods that are surplus to production or not produced locally. Support for this vision in India can be drawn from the writings of Mahatma Gandhi, indigenous peoples' organizations, and some farmers' unions in India and elsewhere.

Each vision was introduced with a 30–minute simulated video news report that sought to highlight key aspects of each possible future scenario and help jury members to visualize their implications.

Expert witnesses. Following the video presentations, expert witnesses presented the case for a particular vision of the future. Members of the AP Government, the corporate sector, and civil society organizations were given equal amounts of time to present their case to the jury. Jury members were allowed to cross question expert witnesses after their presentation.

Jury deliberations. Jury members considered all three visions, assessing pros and cons on the basis of their own knowledge, priorities, and aspirations. The different contributions of invited expert witnesses were important for the jury's deliberations. The jury members were not asked to simply choose between visions one, two or three. Instead, outsider facilitation encouraged them to critically assess the viability and relevance of each scenario for the future. They could choose a particular pre-formed vision OR combine elements of all three futures and derive their own unique vision(s).

An oversight panel. A group of external observers oversaw the jury/scenario workshop process. Their role was to ensure that the process was fair, unprejudiced, trustworthy, and not captured by any interest group.

Video archives. The entire citizen jury/scenario workshop process along with interviews of various actors was documented on digital video to:

- provide a clear and accurate record of the event, including the location, the jury setting, the participants, the nature and quality of the debates, the process, and its outcomes, and

- allow any party or external agency to learn from this experience or check for shortfalls in balance, fairness or failings in the deliberative process.

Media. News and media professionals were invited to the Prajateerpu event to relay information about the jury deliberations and outcomes to a wider audience, both nationally and internationally.

Source: Pimbert and Wakeford, 2003

The coalition of five organizations judged that the *Prajateerpu* process achieved its aims in terms of:

- opening up a political space for marginal smallholder farmers and landless labourers to influence the policies of the AP Government and foreign aid donors;
- allowing people without formal education, often without basic literacy, to analyse issues and directly question specialists concerning future policies affecting their community;
- beginning to build a wider coalition of support for the farmers' vision regionally, nationally, and internationally.

Following the jury, the UK press revealed that DFID's own staff had grave concerns concerning Vision 2020. Subsequently DFID withdrew its support from the agricultural component of Vision 2020 (Pimbert and Wakeford, 2003). However, the Head of DFID India accused the organizers of undertaking a process that was biased towards Vision 3 (see Box 1). They asked the two UK institutions that had published the Prajateerpu report (Pimbert and Wakeford, 2003) to withdraw it on the basis of it process having misrepresented DFID's activities in AP, using a flawed method of recruitment for the jury, and being influenced by the vested interests of the partners.

The *Prajateerpu* coalition's answers to these accusations are detailed elsewhere (Wakeford and Pimbert, 2004). They noted that the methodology of such deliberative processes is always attacked by those whose interests their outcomes undermine. The coalition stood by its conclusion that DFID had failed to attempt to allow marginal farmers in AP to influence its agricultural development programme. It defended the recruitment methodology for jurors as being as rigorous as possible in the context of rural AP, and as having been more balanced than most participatory exercises. As for the accusation of bias, it pointed to the wide range of individuals and organizations that had been part of an oversight panel that had ensured the fairness and competence of the process.

In terms of contributing to policy change, the additional publicity created by the controversy had made a significant contribution to generating public disquiet over the authoritarian modernization plans of AP Chief Minister Chandrababu Naidu (who fell from power in 2004).

2.3 *Izwi neTarisiro*: adaptation to the Zimbabwean context

The principle organizer of the Zimbabwe jury, Elijah Rusike of ITDG Southern Africa, worked closely during 2002 with some members of the *Prajateerpu* team, in order to draw lessons and make modifications to the plans for Zimbabwe.

The citizens' jury process was christened *Izwi neTarisiro* in Shona, which literally means 'People's Voice and Expectations'. The first adaptation was to hold an issue identification workshop, rather than frame the discussions around an existing policy document. The criteria that were agreed for the selection of farmers to the issues identification workshop were:

- the participant had to be a full time farmer;
- full time residence in the rural area;
- there had to be equal gender representation from the same district;
- the participant had to have a broad knowledge of rural area issues;
- an ability to articulate issues.

The organization of a national issues identification workshop as a part of the *Izwi neTarisiro* process was important because, unlike in the *Prajateerpu* process, which could draw on a government-promoted development strategy – Vision 2020 – and develop alternatives to it, the Zimbabwe process took place when normal channels of policy formulation processes had broken down, and officials were unable to offer any detailed long-term vision for the agriculture sector. Political attention was focused overwhelmingly on the controversial land redistribution process, as this statement from a ZANU-PF MP demonstrates.

> *The land question has been and remains the central social, political and economic issue for Zimbabwe today. It is central to the attainment of social spiritual and political stability and economic development in the country. The majority of the nation's poor particularly rural families live directly off the land. The land reforms are an attempt to redress the legacy of colonialism. It cannot be expected therefore that the private sector should play a leading role in delivery of land back to Africans.*
>
> (Ncube, 2003)

Thus government initiatives on the future of agriculture focused upon the immediate priority of speeding up the redistribution of land through a series of measures and decrees broadly under the umbrella of 'Fast Track Land Reform'. A fundamental difference with Andhra Pradesh was that due to political conditions prevailing nationally and internationally around the land question in Zimbabwe, conditions were not right for a public debate on this most high profile issue. Therefore a more generalized vision-focused question was used as the expected output for the jury process. The overarching question for the witnesses and jurors to address during the week-long deliberations was: 'What Do You Desire To See Happen In The Smallholder Agriculture Sector In Zimbabwe By 2020?'. As discussed above, the *Prajateerpu* commenced with three half hour films laying out the three different visions for the future of agriculture. In the Zimbabwe case the jury was not presented with a set of distinct scenarios at the outset, mainly due to a lack of resources to create the films rather than any conscious decision to avoid the accusations of bias.

In order to create the most conducive environment for the poor farmers of Andhra Pradesh to speak out confidently and with maximum impact, the organizers of *Prajateerpu* put forward a number of principles (Pimbert and Wakeford, 2002), not all of which were adhered to in the Zimbabwean case. These are looked at in turn below:

- *The process should reflect the perceptions, priorities, and judgments of ordinary farmers.* The social gulf between senior government officials and ordinary villagers was nowhere near as pronounced in Zimbabwe as in Andhra Pradesh. It was not necessary to engage in a series of processes which would 'put the jury at ease'. The farmers' representatives quickly got into their stride in terms of vigorous cross-examination of specialist witnesses.
- *Holding process in a rural setting.* The *Izwi neTarisiro* was held at an NGO training center in Harare, which meant that it was the farmers who had to travel by bus from all corners of the country. Nevertheless, this was a humble venue and a far cry from the international hotels where policy events are usually held in Zimbabwe.
- *Encouraging bureaucrats, scientists, and specialist witnesses to travel.* Government officials were facing severe fuel constraints and asking them to come to venue in Harare was already quite a major request.
- *TV and video technology – free circulation of information.* Due to the exceptional polarization of media coverage, this was the main and unavoidable weakness of the process in Zimbabwe as far as the *Prajateerpu* principles are concerned. The local media would have been highly critical of the event for stirring up opposition, whilst the international media would have thrown the spotlight on the inadequacies of the government. Both farmers and government officials would have been reluctant to speak out. Nevertheless, the whole of the jury event was recorded on video with a view to the creation of future publicity. The video recordings have been used for creating transcripts in English of the most interesting and crucial stages of the process.

The jury was selected purposively, to achieve a wide representation of districts, ensuring the participation of Ndebele as well as Shona communities. The tendency in Zimbabwe is for Ndebele speakers to understand Shona but not so much the other way round. So the jury itself was conducted in Shona, with bilingual farmers interpreting into Shona when representatives from Matabeleland districts were speaking. The partner NGOs involved in setting up the *Izwi neTarisiro* began a process of jury selection by identifying informal leaders from amongst the farmers they worked with. Most had participated in interventions by NGOs and had their confidence built to interact with outside institutions. They were selected by reputation for being articulate, which was not necessarily synonymous with literacy. Gifted debaters who were known in the communities were brought forward. Youth were not a well-represented group amongst established small-scale farmers – this tended to influence the discussion of topics such as HIV/AIDS.

The national partners to the *Izwi neTarisiro* in Zimbabwe were the Biotechnology Trust of Zimbabwe (BTZ), Community Technology Development Trust (CTDT), BioSafety Board of Zimbabwe, and Scientific Industrial Research and Development Council of Zimbabwe. The partners were coordinated by ITDG Southern Africa. The Institute for Development Studies, University of Sussex, the formal UK research partner in the project was unable to send a representative. Two staff from the ITDG Head Office in the UK, Stuart Coupe and Jon Hellin, agreed to step in and participate as international observers and commentators. Funding was sourced from the Rockefeller Foundation.

Details on the participant farmers, the expert witnesses and the verdicts reached by the farmers on each of the topics debated can be found in the Appendices. In the following sections the reader is introduced to six major policy areas in Zimbabwe and the response of the jury to them is analysed in detail. The first section looks at the headline issues affecting Zimbabwe at this time from an international perspective; land reform and HIV/AIDS. The second reviews interactions between the small-scale farmers and their representatives in national policy dialogue, the Zimbabwe Farmers' Union. It is the area where the jurors felt most informed and empowered to speak out and was also the most revealing in terms of the attitudes of communal area farmers towards established institutional frameworks. The third section of interest is on new technologies and new varieties in the seed sector, a topic of vital importance to the farmers but where their access to knowledge and information about current and future policy developments is greatly constrained.

3
Headline issues

Over five days, the farmers formed a jury to hear evidence from a range of policy-makers on the future of agriculture. In all, they cross-examined witnesses for 11 different priority topics. This paper does not attempt to look at each topic in detail. Instead examples are given of three types of topics:

- Issues at the top of the agenda in terms of local and global policy: land reform and HIV/AIDS.
- An issue that generated the most passionate and heated exchanges of the week: the role of the Zimbabwe Farmers' Union in representing the interests of small-scale farmers.
- Topics that were quite new to most of the farmers, where they considerably advanced their understanding of the policy issues at stake in the process of reaching a verdict.

In the following sections, the recent history and current government policies in the particular area are broadly outlined before examining the perspectives of witnesses and jurors.

3.1 Land reform

The land reform process in Zimbabwe since Independence has involved reducing the number and size of large-scale commercial farms to smaller units operated by individual indigenous farmers and groups of farmers in accordance with different models of resettlement. The country has undergone two broad phases of land acquisition and subsequent resettlement. The first phase, from 1980 to 1998, was characterized by gradual acquisition of land by government and elaborate planning prior to resettlement. The government implemented various different resettlement models, the most dominant being Model A, with small plots of three to five hectares allocated either in a village pattern with communal infrastructure or self contained farming units. Schemes were managed by Resettlement Officers with substantial support from different ministries. The second phase from 2000, Fast Track, due to inadequate resource availability and its accelerated nature, provided for land allocation and resettlement without detailed planning and infrastructural provision (Derman and Gonese, 2003).

At the start of the fast track resettlement in 2000, land occupations were clearly an illegal process according to the legal framework that applied at the time. This included the Constitution of Zimbabwe, Land Acquisition Act, and government policy on illegal settlements, which all recognized the supremacy of private property rights. The government then embarked on a total revamp of the legal and judicial framework in an effort to normalize the situation created by occupation and fast track resettlement that had thrown the concept of security into disarray, particularly as it relates to freehold lands.

Farm occupations and the fast track resettlement that followed created an environment of uncertainty with regards to the land rights of the affected large-scale farmers while those of the incoming settlers remained largely unclear and unprocessed. The legitimization of land occupation by the Rural Land Occupiers Act has made freehold title for rural land one of the most insecure forms of land tenure (Marongwe, 2003).

Box 2   A resettlement experience – Midlands

Zimbabwe's fast-track land reform programme is meant to benefit landless people forced to live in congested communal areas, but many of the supposed beneficiaries are turning their backs on their new land.

The Dorset resettlement area, 40 km south of the Midlands capital of Gweru, is an example of the shortcomings of the accelerated land redistribution programme – which was meant to reverse the legacy of a century of colonial land policy.

'I went to Dorset in 2001, during the height of farm invasions. At first I was sceptical about Kujambanja, but when I saw a significant number of my neighbours leaving, I decided to join the trek', Furusa told IRIN.

During the early days of the fast-track programme there had been a sense of euphoria 'about farm invasions, and I genuinely believed that, at last, I would be a proud owner of my own piece of land'.

'[But] I discovered that the area we had been made to move into did not have good soils, having been reserved for cattle ranching. In addition to last year's insufficient rains, there is no way in which the new farmers there could get good harvests owing to the poor soils, which are just as bad as where I come from', added Furusa.

Since he had only two head of cattle for draught power, he said, preparing his plot was proving too difficult – a situation that left him with no option but to return to his father's home, where he could pool resources with his extended family.

The father of three charged that by moving thousands of people to unsuitable land, veterans of Zimbabwe's liberation war and the government were only interested in getting their votes in the parliamentary and presidential elections (in 2000 and 2002, respectively).

Like the other settlers turning their backs on Dorset, Furusa complained that schools were very remote and it would be difficult for his two school-going children to travel the distance. The Dorset resettlement area also lacks proper health facilities, and transport is mostly by ox-drawn cart.

Furusa said a significant number of the land occupiers who had moved to Dorset and nearby farms were resorting to gold panning in the Mutevekwi River, which runs close to the small town of Shurugwi, in order to survive.

Source: Irinnews.org, 4 September 2003

Official policy statements on beneficiary selection for the resettlement programmes provide ample opportunity for the participation of a wide range of stakeholder institutions. Empirical observation suggests a diversity of operational departures from policy provisions, given the politicization of the land issue against the background of the 2000 and 2002 Parliamentary and Presidential elections. Numerous decisions by respective Identification and Selection Committees have reportedly been ignored and/or reversed by political heavyweights whose actions not only compromise formal planning processes, but also seriously undermine the role and effectiveness of officials responsible for selection of beneficiaries for Model A1 and A2 schemes (Gonese and Mukora, 2003). As one senior government official admits, 'the selection and allocation of settlers was often highjacked by influential groups or individuals, resulting in some settlers being thrown off the plots, or violent confrontations' (Mundeiri, 2003). There is also some evidence that poor small-scale farmers who did respond to the rush for land, but lacked political influence tended to be allocated less valuable lands with poor or degraded soils, see Box 2.

The most explicit discussion on the land reform process occurred following a presentation by the Department of Agricultural Research and Extension (AREX). Whilst many countries have privatized or run down their state extension service, the policy of the Zimbabwean government is to maintain such services. They are nevertheless severely under resourced, leading to prioritization of certain categories of farmers for support:

DIRECTOR, AREX: *I have to clarify also that part of the reason why we fail to reach out to everybody is our current policy, which tends to give more priority to resettlement farmers rather than communal farmers.*

FARMER: *It seems to me that there is no proper balance between support to resettlement and communal area farmers.*

FARMER: *The education system is replete with incentives based on success, reward and freedom to move to greener pastures on graduation. What about us farmers? Am I expected to rot in one place in spite of having attained higher farming skills and proven experience?*

DIRECTOR, AREX: *You may be the odd one out, but the entire land resettlement programme, which I happen to have helped design and formulate, was aimed at farmers like you. When the programme gets evaluated soon I bet you will find that the majority of those allocated land were experienced communal farmers.*

FARMER: *What chances exist for a review of the current resettlement programme so that land allocated to inexperienced farmers can be retrieved and reallocated to more competent communal farmers?*

As we can see from this exchange, officials participating in the jury process expected farmers to be excited by fast track land reform, and were unsettled by critical questioning. The perceived benefits of redistribution appeared to be undermined by processes of beneficiary selection lacking in transparency. Land reform is a process of great magnitude, involving migration to a new area, but the risks are great. The latest wave has become a very individualistic process, with local leaders like kraal heads not given any role in selecting beneficiaries. Land seizure on commercial farms have displaced farm labourers and created an atmosphere of violence and unrest which communal area farming communities, used to living peacefully, found threatening and disruptive. For the communal area farmers there is a crisis of expectations: the process is going against all their expectations that good farming knowledge and skills would be recognized and rewarded. None of the farmers on the jury had moved to the new resettlement areas, but they were highly aware of the risks involved. Therefore, it is possible for us to lay out the position of communal area farmers: that the current resettlement process is leading to the destruction of agricultural productivity and is detrimental to the future of agriculture in Zimbabwe.

3.2   HIV/AIDS

There are an estimated 40 million cases of HIV/AIDS worldwide and 95 per cent of these are in developing countries (UNAIDS, 2004). Southern Africa is home to the worst HIV/AIDS epidemic in the world. Almost 15 million people in southern Africa were living with HIV/AIDS at the end of 2001 and an estimated 1.1 million died of AIDS in 2002, the majority of them in their productive prime (UNAIDS, 2003).

Seven countries in the region – Botswana, Lesotho, Namibia, South Africa, Swaziland, Zambia, and Zimbabwe – are reported to have adult HIV prevalence of more than 20 per cent. According to UNICEF, in Zimbabwe 600 000 children have been orphaned by HIV/AIDS and some 2.2 million (30 per cent of the adult population) are living with HIV/AIDS (see Table 1). The impact that HIV/AIDS will have in Africa is being likened to the bubonic plague in Europe in the fourteenth century when populations in villages and regions were reduced by two-thirds (Runge et al., 2003: 24).

There are a number of reasons for the rapid spread and high prevalence of HIV/AIDS in Southern Africa. These include poverty and economic marginalization, poor nutrition, opportunistic infection, migration, patterns of sexual contact, and gender inequality.

Table 1 HIV/AIDS prevalence in selected Southern Africa countries

	Estimated number of people living with HIV/AIDS in 2001			New AIDS orphans 2001	New AIDS deaths 2001
	Total	Adults	Adult rate		
Zimbabwe	2 300 000	2 000 000	33.7%	780 000	200 000
Zambia	1 200 000	1 000 000	21.5%	670 000	120 000
Mozambique	1 100 000	1 000 000	13.0%	420 000	60 000
Malawi	850 000	780 000	15.0%	470 000	80 000
Lesotho	360 000	330 000	31.0%	73 000	25 000
Swaziland	170 000	150 000	33.4%	35 000	12 000
Total	**5 980 000**	**5 260 000**	**20.4%**	**2 448 000**	**497 000**

Source: UNAIDS, 2002

Impact of HIV/AIDS at the macro-economic level

One of the United Nations Millennium Development Goals is to combat diseases such as HIV/AIDS (The World Bank Group, 2004). Without making substantial headway, however, in preventing and mitigating HIV/AIDS, it is highly unlikely that the other goals such as the halving of poverty and hunger will be met. This is largely because HIV/AIDS distorts spending and eats away at savings.

On average, HIV/AIDS-related expenses can absorb one-third of a household's monthly income. People spend more money on health care, traditional ceremonies, and funerals. Funeral attendance is an obligatory custom in many parts of rural Zimbabwe. In 2003 the average expenditure on funerals in Zimbabwe was US$105 (The Economist, 2004). The farmers' jury pointed out that attendance at funerals and social gatherings such as *kurova guva* (a function held one week after someone's death) was having a detrimental impact on farming operations. Furthermore, people with HIV/AIDS are often too sick to work in the six months before death.

Despite the obvious detrimental impact of HIV/AIDS, attempts to assess this impact at the macro-economic level are frustrated by the fact that many factors affect economic performance. However, it is estimated that HIV/AIDS is already depressing sub-Saharan Africa annual gross domestic product (GDP) growth rate by approximately 0.8 percentage points. In the worst hit countries such as Zimbabwe, where more than 20 per cent of adults have HIV/AIDS, the figure is approximately 2.6 percentage points. The impacts of HIV/AIDS are a great deal easier to identify for specific sectors and below we consider the impact on the agricultural sector.

The challenges facing smallholder farmers in Zimbabwe and the rest of Southern Africa have been hugely accentuated by the impact of HIV/AIDS. One of the principal reasons is that labour is a very important component of agricultural production in Sub-Saharan Africa because of the limited use of purchased inputs. Labour shortages due to HIV/AIDS are having a very detrimental impact on the agricultural sector.

HIV/AIDS and the agricultural sector

The impact of HIV/AIDS in terms of morbidity and mortality is particularly severe in the agricultural sector. According to the FAO (2001) some 7 million farmers and farm workers in 25 African countries had died of AIDS by 2000 and that 16 million more will die by 2020. The FAO also anticipates that in the most affected countries, HIV/AIDS is going to reduce the agricultural labour force from 10–26 per cent. The figure for Zimbabwe is 22.7 per cent (see Table 2).

Research conducted by the United Nations Development Programme found that the HIV/AIDS prevalence rate on farms in Zimbabwe was 43 per cent, with the highest number of HIV-positive people in the 15 to 23 age range. This age group represents the core of the agricultural labour force and should be the most productive economically.

Table 2 Impact of HIV/AIDS on the agricultural labour force in the most affected countries (projected losses in percentages)

Country	2000	2020
Namibia	3.0	26.0
Botswana	6.6	23.2
Zimbabwe	**9.6**	**22.7**
Mozambique	2.3	20.0
South Africa	3.9	19.9
Kenya	3.9	16.8
Malawi	5.8	13.8
Uganda	12.8	13.7
Tanzania	5.8	12.7
Central Africa Republic	6.3	12.6
Ivory Coast	5.6	11.4
Cameroon	2.9	10.7

Source: FAO, 2001

In Zimbabwe, the large-scale farming sector is the major course of food production. HIV/AIDS affects commercial farms in the same way as it affects any other business. For the farm owners, HIV/AIDS leads to higher costs in terms of paying employee benefits, staff turnover, and recruitment and training.

Rates of HIV/AIDS are higher in rural areas because information and health services are less available in these areas than in cities. In addition, rural people are less likely to know how to protect themselves from HIV. Furthermore, if they fall ill they are less likely to receive adequate care than in urban areas. One of the jurors summed up the difficulty of containing the spread of HIV/AIDS:

> *I understand that traditionally people infected with sexually transmitted diseases were bound up with ropes, forced into sacks and then taken outside the village for burning. This was not only a way of instantly stopping the spread of the disease but a stern warning to those potentially involved in sexual immorality in future. However in the case of HIV/AIDS, there is so much secrecy about those infected let alone HIV/AIDS related deaths. In my opinion this is a major factor contributing to the spread of the disease and I am at a loss concerning the way forward.*

Another jury member commented that:

> *Most of us small-scale farmers lack education and our leaders look down upon us. They don't pass vital information on HIV/AIDS to us. We therefore advocate for the introduction of programmes that bring awareness directly to those at the grassroots.*

A woman jury member pointed out the link between gender inequalities and the spread of HIV/AIDS:

> *We are aware that this problem started in the cities where our spouses have gone in search for employment. The tragedy for us rural women is that when they come back they insist on unsafe sex without condoms. When we refuse they are quick to accuse us of having other sexual partners.*

The high rate of infection among women is having and will continue to have enormous implications on nutrition and food security. Women head many farm households and in Sub-Saharan Africa women often provide a large portion of the total labour force. HIV/AIDS is, therefore, having a huge impact on food security in the region.

Impact on food security

Following the World Food Summit in November 1996, The Rome Declaration on World Food Security was issued. Food security was defined as 'food that is available at all times, to which all persons have means of access, that is nutritionally adequate in terms of quantity, quality and variety, and is acceptable within the given culture'.

Households are able to achieve food security when they can produce sufficient amounts of nutritious food, earn enough cash income to purchase food, sell or barter assets for food in hard times, and rely on social support networks for assistance. The HIV/AIDS epidemic is eroding each of these coping methods. In order to meet the costs of HIV/AIDS medical costs, families may have to use their savings, sell assets such as land and livestock, borrow money, or seek support from their extended family. HIV/AIDS, therefore, reduces households' capacities to produce and purchase food, depletes their assets, and exhausts social safety nets (UNAIDS, 2003).

There is now a danger that Southern Africa has crossed a threshold and that, irrespective of climatic conditions, food insecurity will last for years because of the debilitating impact of HIV/AIDS. In some cases agricultural output can decline by 50 per cent in farm households affected by HIV/AIDS. In Zimbabwe, the production of maize, cotton, sunflowers, and groundnuts has been particularly affected.

Remote fields tend to be left in fallow and, in general, yields decline as a result of delays in essential farming operations, lack of resources to purchase agricultural inputs, and the abandonment of soil conservation measures. Also as people fall ill, agricultural educational and extension services may be disrupted. In addition, children may be taken out of school to fill labour and income gaps created when adults fall ill and/or start caring for family members with HIV/AIDS.

HIV/AIDS in rural areas can lead to the loss of many grains, tubers, legumes, and vegetables and a decline in the nutritional quality of people's diets. This is because farmers, facing labour shortages, reduce the area under cultivation and reduce the number of crops that they grow. HIV/AIDS farm households often switch from labour-intensive export crops to less labour-demanding subsistence crops (see Table 3).

Table 3 Labour demands of selected crops

High labour intensive farming	Low labour intensive farming
Vegetables	Cassava
Sesame	Sweet potato
Groundnuts	Pulses
Cotton	Yams
Tobacco	Sorghum

Source: SADC/FANR, 2003

Toupouzis (2003) has reported that, as result of HIV/AIDS, in the Bukoba District in Tanzania, the intensely managed banana/coffee/bean agricultural system has been largely replaced by a less labour-demanding cassava/sweet potato system. The change in the cropping system is unlikely to reverse itself because of continued labour-shortages due to HIV/AIDS.

Farm families can get caught up in a vicious circle: the lack of sufficient food and adequate nutrition is particularly detrimental for the health and well-being of people living with HIV/AIDS. Malnutrition weakens the immune system and can, therefore, lead to the accelerated development of AIDS-related illnesses in HIV-positive people. Furthermore, people living with HIV/AIDS have increased nutritional requirements, especially protein. Hence, while a crop such as cassava may be a good choice to grow in areas where there are severe labour shortages, it has a low protein concentration and is unable to meet the protein needs of HIV/AIDS sufferers.

Another worrying consequence is that local knowledge about seeds and crop varieties are lost because HIV/AIDS-infected adults not only stop planting these varieties but also fail to pass on the valuable indigenous knowledge to the next generation. HIV/AIDS deaths can also have an adverse impact on post-harvest activities such as food storage and processing, as well as seed storage for the next year's cropping period.

Labour shortages due to HIV/AIDS can lead to failures to look after livestock. Livestock serve multiple functions in many rural areas, including being a source of food, traction, and fertilizer. Increased thefts of cattle are often a result of a failure to herd the animals, while cattle are exposed to tick-borne diseases when they are not dipped properly. Livestock often have to be sold to pay for the medical costs of HIV/AIDS and funeral rites may involve the slaughter of animals.

Men are usually responsible for marketing agriculture produce and when they die, widow-headed households suffer from reduced market opportunities for crops such as maize and cotton. There are many parts of Sub-Saharan Africa when women lose all their household assets when their husband dies.

Caring for sick relatives

An increase in the number of HIV/AIDS sufferers and also the number of orphans caused by HIV/AIDS deaths enhances the child-care responsibilities of predominantly healthy women in the farming community. In general, the costs of HIV/AIDS are largely borne by rural communities because HIV-infected urban dwellers of rural origin often return to their rural communities when they fall ill.

By 2010 the Southern African region is expected to have around 5.5 million maternal or double orphans, approximately 16 per cent of all children under the age of 15 years, of which 87 per cent will be orphaned because of the HIV/AIDS epidemic. Care of sick relatives was a concern for the farmers' jury: after the presentation about the HIV/AIDS situation in Zimbabwe a woman juror commented that:

> *It is crystal clear from the presentation that the trend in the spread of HIV/AIDS is increasing at an alarming rate. We are aware that in response to this, government has set up the National Aids Council, amongst other initiatives, whose mandate is to mitigate the problem through a home based care programme. However, it has come to light that the home-based care programme is disrupting farm productivity because farmers are now spending precious time caring for patients.*

The same woman juror went on to ask if the researchers had ever approached the government with recommendations for alternative solutions such as the establishment of community centres where all patients can be taken care of by people specifically employed for this job. The researchers' response was non-committal:

> *We have not made such recommendations. On the contrary, we have information that home-based care programmes are more successful because relatives tend to provide more love and care to the patients. We would rather go with a suggestion that more funds to be channelled toward employing more full time carers within a home based care programme. Considering the magnitude of the problem at hand vis-à-vis the resources available, it is more credible to recommend that government should influence more donors to support the home-based care programme.*

It is debatable, however, whether donors will be prepared to support such a programme partly because of the acrimonious relationship between the donor community and the government and partly because the numbers requiring care is set to increase. There is also the issue of corruption: another jury member raised his concerns about the misappropriation of funds:

> *In my district there is stigma associated with HIV/AIDS related deaths leading to people being given indecent burial. Although government has come up with a fund to assist with HIV/AIDS related burials, the fund has been misappropriated by the community leaders. There is also a lot of investment in raising the awareness of the leaders on HIV/AIDS yet the benefits never trickle down to those directly in need. I therefore recommend that this training be targeted on community members rather than the leadership.*

Controlling the spread of HIV/AIDS

The Farmers' Jury was of little doubt that HIV/AIDS poses a real and growing threat to rural people's livelihoods. It considered the challenge of how best to control the spread of HIV/AIDS in the context of a society where open discussion about sexual matters is taboo. Hunger is forcing many people into increasingly high-risk survival strategies. As people become desperate for food and other resources, they might migrate to urban areas in search of employment or engage in prostitution, thereby increasing their vulnerability to HIV infection. There are indications that food shortages are driving more women and girls to transactional sex, whether for cash or for food, in the six countries of Southern Africa affected by the crisis. Furthermore, while better transport facilities help farmers to market food surpluses they may also increase travel-related HIV susceptibility. As one jury member pointed out:

In our culture it was common for the youths to receive sound counseling and advice on life and sexual matters from village councils of elders for boys and for girls. This is contrast to the youths of today who rely on the contemporary educational system that cannot deliver on this subject. I also believe poverty is one of the main causes of HIV/AIDS. For instance without income generating projects, carers of HIV/AIDS orphans especially first born girls in such families, feel pressured to engage in sex trade as a livelihood option.

The researchers pointed out that:

Uganda seems to have made tremendous progress in mitigating the spread of the pandemic. This is because its people were quick to take some corrective measures. This is unlike countries in Southern Africa. We should therefore strive to learn from the Ugandans. However, the best remedy is prevention rather than cure.

Some members of the Farmers' Jury adopted a strict moral attitude vis-à-vis the spread of HIV/AIDS:

Everybody fully appreciates the extent to which HIV/AIDS is severely affecting our society particularly the youths. In my opinion this is a direct result of introduction of the Age of Majority Act by government. This has given our youths above 18 years of age a free passport to do as they wish. This is goes against traditional culture whereby youths were expected to be dependent on parents even into their 30s or 60s. Given the way these youths indulge in beer drinking I am not surprised the incidence of HIV/AIDS is high in this age group.

Other comments from the jury included:

My opinion is that all people should stick to one sexual partner. Secondly, I support an initiative to invest in the production of nutritious foods by small-scale farmers. Thirdly, it would make sense to secure an in-depth understanding of the spread of HIV/AIDS by regions including trends and root causes. Fourthly I think there is a weakness in the focus and approaches of all the aids action committees in our districts. Rather than focusing on improving awareness among potential HIV/AIDS victims, all these committees concentrate on counselling of those infected. I grew up in a mine compound and was struck by the wisdom of mine management who introduced a rule that all female visitors from the cities were to be tested for sexually transmitted diseases. They knew that all their workers could easily become victims. Similarly, in the case of HIV/AIDS, there is a great sense of urgency in testing people to establish their status.

HIV/AIDS is a highly emotive subject for poor farmers. It is a most devastating pandemic with unprecedented social, economic, and cultural impacts. Major policy changes are needed to tackle multiple confusions surrounding HIV/AIDS and the failure of existing initiatives to provide support to the needy.

4

Farmer representation

The most dynamic and passionate sessions of the *Izwi neTarisiro* were the appearance of senior national representatives of the Zimbabwe Farmers' Union. Whilst the tone of the cross-questioning of government officials was politely sceptical, ZFU witnesses were subjected to bitter criticism. With ZFU being the official representative of small-scale farmers in national policy processes, it was an opportunity to raise important issues of accountability to ordinary members.

4.1   Zimbabwe Farmers' Union

The following assessment of ZFU was conducted in the late 1990s, before the economic crisis of recent years. It serves to illustrate how ZFU was operating at its fullest capacity.

Box 3   Zimbabwe Farmers Union

ZFU was formed in 1991 from the merger of the Zimbabwe National Farmers Union (ZNFU), representing the small-scale commercial farmers, and the National Farmers Association of Zimbabwe (NFAZ), representing communal area farmers.

Structures
There are a number of different levels of organization in the ZFU, with some diversity inherited from the base structures of the two parent unions:

Base – clubs (over 6 000 with typical membership of 15–40).
Area – area councils representing clubs and commodity groups often involving slightly larger scale farmers.
District – District ZFU Council. With paid official(s) in a majority of districts.
Province – A Provincial Committee, commodity committees, and staffed office.
National – Annual Congress, National Council, Sub-committees, commodity committees, and paid staff.

ZFU has a very diverse membership that can be broken down in a number of different ways:

Farmer type – Communal area, resettlement area, small scale commercial, large scale commercial (indigenous), and peri-urban plot holders

Membership type
Active – have paid membership fee to support work of ZFU and/or are involved in various ZFU committees or activities. They have the right to vote.
Nominal – have paid membership fee purely to make use of sales tax advantages, no other involvement. They have the right to vote.

Inactive – affiliated to ZFU clubs but have not paid fees. They have no right to vote.
All other smallholders – not technically members, but some feel ZFU represents their interests; at national level ZFU is often thought to represent smallholders in general and the levy income of ZFU comes from all smallholder sales, so they all contribute indirectly to ZFU.

Activities
ZFU has an extraordinarily wide range of activities, many going beyond what are traditionally seen as union activities:

Lobbying – supported by the Economics and Planning Department, ZFU does research and makes representations on a wide range of issues relating to agricultural policy and prices.
Sometimes this is done in conjunction with the other farming unions and sometimes independently.
Research and extension policy – this is a more specific form of lobbying in which ZFU has a place on the Agricultural Research Council and similar bodies. In addition ZFU has used donor funds to support specific pieces of research.
Extension Access – ZFU groups are active at a local level to ensure they and their members get access to official advice and support.
Inputs – ZFU, particularly at District Office level, is involved in bulking up input orders and negotiating volume discounts from suppliers.
Marketing – ZFU supports the development of farmer marketing collaboration at the local level, putting ZFU groups in contact with potential purchasers. Provision of market information through a weekly radio programme (Shona and Ndebele), monthly magazine (mainly English with some Shona and Ndebele), and an occasional marketing news sheet (currently in English).
Training – both in organizational issues such as leadership training and gender awareness and technical training such as the production of technical manuals.
Commercial – ZFU has a transport fleet which it operates on a commercial basis, but with an emphasis on serving communal areas. ZFU also has share holdings in a number of organizations. ZFU policies and activities are orientated towards the commercialization of smallholder farming; this seems very similar to the objective of Agritex and the Ministry of Agriculture.

Finances
There are three main sources of income for ZFU:

Levies – representing nearly half of income – these come from a 1 per cent levy on communal/resettlement area sales. This has been complicated by the recent liberalization of marketing, giving more channels and less division between communal and large scale commercial areas. GMB is still paying the levy but may ask for an administrative fee for deducting it,
Grants – approximately 40 per cent of income, mainly for specific projects supported by donors.
Membership fees – 10–20 per cent of income.
Source: World Bank, Rural Producers Organisations, 1999

Up to the year 2000, ZFU had managed to gain strong political influence through its institutionalized relationship with government. ZFU is recognized by government as the legitimate representative of communal, resettlement, and small-scale commercial interests. Recently, the organization's scope to retain the freedom to disagree with and challenge government policies has diminished considerably. ZFU's ability to speak for smallholder farmers is also complicated by the fact that only one-fifth of its potential constituency are paid-up members. ZFU as a whole has not yet had much influence on AREX priorities or strategies at the national and provincial levels. Equally, it has had relatively little influence over national research priorities.

From 2000 onwards, the impact of declining government revenues, donor boycott, and immiseration of farmers has had a devastating impact in ZFU. Certain European donor agencies backed ZFU for many years. The union became less dependent on members' dues. But national ZFU leaders were drawn into politics, both government and opposition – reducing their commitment to grassroots issues. Then when donors saw the Union as becoming politicized, they withdrew support – and now it is virtually collapsing under resource constraints. To take a case from ITDG field experience: the Mutare ZFU office is very small, with no transport, and recently the telephone was disconnected. The District Coordinator for Nyanga is based in Mutare, 105 km away from the ITDG project area. The ZFU local coordinator faces acute constraints in communicating with members there. ITDG participated in a restructuring of ZFU at the grassroots level. Commodity associations were established in Nyanga to bring focus to ZFU support on the marketing of produce. Unfortunately it has not been possible for the ZFU to give systematic support to this initiative.

The inability of ZFU to maintain good coverage in support and services to its members set the stage for some sharp exchanges between the Farmers' Jury and the Director of ZFU.

FARMER: *From your presentation I understand that ZFU is a well-established organisation founded a long time ago. You also mentioned that the organisation has representation in all districts across the country and that your mandate is to support small-scale farmers through training, amongst other things. However, whilst I have heard about this union, I will be very honest and say that I still haven't seen the difference ZFU has made to farmers. All I know is that small-scale farmers have been paying membership fees. As for the projects you have been speaking so articulately about, I am yet to see.*

DIRECTOR, ZFU: *One of the problems the Union faces is lack of transport for district leaders to visit farmers to disseminate information and raise awareness on projects agreed upon by ZFU at national level. Secondly, some of the leaders elected by members are not very committed or enthusiastic about representing farmers. This is why ZFU and its partners such as ITDG decided to launch a capacity building programme aimed at helping farmers to find solutions to issues affecting their livelihoods through discussion in appropriate forums.*

FARMER: *I am afraid, I will have to speak like one highly critical and offending you. This is because I am a ZFU member with all the information about the union. When we formed this organization at independence, I thought all our problems were to be eradicated. But I have just learnt from your presentation, this may well take another 22 years. I can tell you that our government is fully behind you hence why they permitted us to form the Union in the first place. Now for as long as you speak the way you do, ZFU will never be progressive. How can you dare to say ZFU cannot meet small-scale farmers because there is no transport, and yet when we formed this union we didn't have any funds let alone transport? If it means the leaders have to walk long distances to meet the farmers let them do so. If you have been chosen as the leaders to represent people, then you need to be committed to your duties. Consider this classic example. We did not establish ITDG, yet it is ITDG that has gathered us here today. We therefore need to be informed about what is happening at national level, otherwise we will not appreciate why you don't want to see us and what you stand for. Secondly, I want to challenge you to tell us how many communal farmers ZFU has represented to be allocated farms under the resettlement programme?*

COMMENT FROM FACILITATOR: *It looks like we are all focusing on the negative things. Surely there must be something good about ZFU?*

FARMER: *I insist that ZFU's problem is not at district but national level. I remember vividly what happened way back before independence. The farmers' association then used to collect a very small levy over many years from farmers. Then in 1969 a serious drought occurred and the association responded by requesting land development officers to go around asking all the farmers to list down their needs. That year, farmers received large sums of money from the association with most being spent on capital items such as scotch carts. This is in sharp contrast to the union we have at present that is after farmers' money. It condones and thrives on favouritism causing problems for its district officials who are otherwise doing their job well.*

FARMER: *You are supporting your cronies in resettlement farms at our expense and using your district officers as scapegoats. We want to know your plans that will help farmers realize their dreams.*

FARMER: *I will not be surprised to see your cars being stoned by angry farmers in future because we feel that we are being exploited and cheated. For example, if I were to receive a telephone call from farmers in my district at the moment, what am I going to report back regarding their future? As a leader and their representative I need a convincing answer for their problems.*

From these exchanges it is clear that the small-scale farmers on the jury still have very strong expectations that their union should function effectively to represent their interests at all levels. There is clearly anger that this is not happening, but the farmers tended to attribute the problem to individual failings from national officials rather than general institutional malaise.

4.2 ZFU, credit, and marketing

There is plenty of evidence that market access is critical to the security of the livelihood of smallholder farmers in Africa. As IFAD (2001: 162) notes 'liberalization of domestic and international markets gives the poor new opportunities for specializing in, and exchanging, their labour-intensive products'. The challenge is that resource-poor farmers seldom understand how the market works or why prices fluctuate. They have little or no information on market conditions, prices and quality of goods; they are not organized collectively; and they have limited experience of market negotiation and little appreciation of their capacity to influence the terms and conditions upon which they engage with the market (IFAD, 2001: 170). One of Africa's main development challenges is, therefore, the delivery of agricultural services (marketing, input supply, financing, and other support) (Coulter et al., 1999; Stringfellow et al., 1997).

In Zimbabwe, access to agricultural commodity markets is still regulated by the state and farmers' access to information, credit, and markets is channelled through agricultural extension services and the Zimbabwe Farmers' Union. Nevertheless the capacity of ZFU is declining, giving rise to further disquiet from the jurors. The most entrepreneurial communal area farmers amongst the jurors, active ZFU grassroots members who are attempting to follow the logic of the ZFU message and respond to cash crop opportunities, are the ones who expressed the greatest disappointment and disillusionment to their national representatives.

> ZFU REPRESENTATIVE: *Having realized that we don't have adequate resources, we embraced a strategy to start pilot demonstration projects in at least one district in all the provinces throughout the country. In these districts computers have been set up to train farmers on use of e-mails and the Internet. This is initially intended for tobacco farmers so that they can check auction prices on a daily basis. While in the past we relied on community structures for information delivery, this has not been efficient, hence the need to assist farmers through the use of modern technology. We also have a monthly newsletter to inform farmers on other issues affecting their industry.*

> FARMER: *Does ZFU ever assess whether the prices farmers obtain from the market compare favorably against costs incurred? This is essential if ZFU are to advise those responsible for setting commodity prices to do so with a conscience. You mentioned that farmers could collude and stop selling their commodities for a certain time until prices firmed up. In 2001, you [ZFU] came to Dande district and advised us to do exactly that and we paid heed. Now we are very upset that you never returned to tell us when we could start selling our cotton lint, despite a promise to do so. We only started selling when we heard through the grapevine that markets were difficult to find and that if we did not sell then, we would lose everything we had.*

> FARMER: *You have an impression that we farmers are not united but I wish to let you know that this is not true. The problem is that ZFU does not want to go into the field to meet the farmers. In Tsholotsho where I come from, farmers perceived an opportunity to grow cotton because of the semi-dry climatic conditions there. The farmers decided to approach the Cotton Company of Zimbabwe (Cottco) Pvt. Ltd. for inputs and implements. They were turned back because they did not have ZFU membership cards. After presenting their case to ZFU they were promised that cards would be issued, but nothing happened. We are disappointed because we could not continue with our plans to grow cotton.*

> FARMER: *We have a long history of working with ZFU. However, your assistance has fallen short of our expectation over time. For instance we feel that you should be obtaining credit application forms for us and help us to complete them. You should also go further and make recommendations to credit institutions on our behalf so that we can obtain loans on the basis of the knowledge that you have of us as individual farmers. In addition, some of us tea and coffee growers had a problem which we presented to our commodity brokers expecting them to solve it for us, but we were surprised when the problem was referred to district and national authorities who never came back to us. So where is the help we are expecting?*

It was clear from the above session on credit and markets that small-scale farmers have not yet become accustomed to seeking out their own market opportunities as has happened in other developing countries where central state marketing and credit facilities have been eliminated. One witness from an NGO, SAFIRE, which has focused on agro-processing using locally available fruits, generated a very positive response from the farmers, who seemed trapped into declining maize and other cash crop producing systems through lack of information and exposure to any potential alternatives. Such alternatives generated a very positive response because they gave farmers a vision on how to expand into niche product markets not controlled by commodity marketing systems. Here are some examples:

I wish to thank the presenter for making me aware that I can generate income from natural resources readily available to me such as mauyu [a gourd like fruit with edible pulp from the baobab tree].

I was delighted when she talked about makoni tea and other indigenous tea species such as zumbani. These are bushes that grow in the wild and I have occasionally picked the leaves to make tea. My question is that is there any technology that can be used to dry the leaves for commercial purpose?

We are so glad and overjoyed because of the information presented. I can only imagine what the presenter must be thinking – that, how is it that these people look so malnourished and skinny when they are sitting on such treasure! All we need now is to have this information on the doorsteps of every farmer because our eyes have not yet been fully opened.

We are so grateful to the presenter for raising our awareness on natural resources management. In my area there are a lot of masau fruit but we have always given them away cheaply to people from the urban areas. Now that I am aware of their real value I will go back and start treasuring them as gold. We should in fact pass a law to regulate how they should be harvested. Just to show you how ignorant I was, I used to weed them out of my fields, thinking they were of no value. Please come down and see us in Muzarabani district.

The process of disillusionment with existing systems of representation and access to marketing and credit services is in full swing, but at the same time the farmers have not yet been able to generate their own collective responses to the problem. None of the potential alternatives open to the farmers, to reinvigorate the ZFU at the grassroots level, or to form a break away union were given any airing. The decline of ZFU capacity has been so rapid that it is still to early to expect any organized reaction by the farmers to the weakness of the organization representing them.

5

Farmer learning

5.1 Genetically modified organisms

The issue of genetically modified organisms (GMOs) became a topic of interest in the *Izwi neTarisiro* process as smallholder farmers felt their voice and opinion had to be heard. Obviously this technology would affect the long-term outlook for smallholder agriculture in Zimbabwe. Despite the fact that most of the jurors did not have much prior knowledge of GMOs, there was lively and interesting cross-examination of specialists after their presentations. The information presented and the subsequent discussion resulted in the formulation of a verdict. Only two jurors had previous exposure on GMOs – they had participated in a visit to Makhatini in South Africa to see Bt cotton trials.

In Zimbabwe GMOs have been promoted by both public and private sector companies, particularly by Monsanto, which has through sponsored field trips, videos, and presentations on the experience of smallholder Bt cotton production on the Makhatini flatlands in KwaZulu Natal, South Africa. Nevertheless Zimbabwean farmers would prefer to see the Bt products performing in their own context than to read about a South African experience. Civil society institutions like Biotechnology Trust of Zimbabwe have facilitated a number of discussion forums and awareness raising workshops on genetically modified organisms as well as conventional breeding and trials of new technologies. It was interesting to note that the usual concerns about GMOs that are raised regularly in discussion on the issues in other regions of the world were also raised by these smallholder jurors.

Background on genetically engineered plants in Zimbabwe

The responsible body for GM policy in Zimbabwe is the BioSafety Board. It is makes its decisions in a context of two broad and opposing visions on the subject.

Industry players who see a commercial opportunity in Bt products argue for a narrow form of risk assessment. They argue that the type of biotech applications being offered are a known quantity and have been widely tested in the US and other parts of the world without problems and have offered substantial benefits to farmers. The chance to increase yields and reduce pesticide applications for key crops such as cotton should not be missed through insistence on unnecessary costly, cumbersome, and lengthy regulatory procedures. Missing out on biotech would in turn undermine the potentials of the seed industry in Zimbabwe and the viability of agriculture in globally competitive markets. The widespread commercialization of Bt maize and cotton in South Africa, and the apparent enthusiasm for biotech products in Kenya, show the benefits and suggest that Zimbabwe is already being left behind (Keeley and Scoones, 2003).

However, others are more cautious. Maize and cotton are central to the economy, and support numerous livelihoods. The export value of both crops may be substantially in their being GM-free, with import niche markets in Europe, and even South Africa (e.g. for maize for baby food). A broader assessment that explicitly incorporates socio-economic, trade, and livelihoods criteria is required. GM crops, it is argued, are linked to a particular style of agriculture that is potentially highly damaging either for specific commodity producers or the most vulnerable farmers. If, for instance, a biotech future is dependent on licensing deals with MNCs this may mean that the costs of inputs go up and the range of technology options is narrowed in a way that is problematic for the most risk-prone farmers who either cannot afford the costs, or ride out the difficulties if promised high returns are not realized. Also the opportunity costs of going the transgenic route may not be the most appropriate, given other available (bio)technologies (Keeley and Scoones, 2003).

Prior to 2001 there were a number of proposals submitted to the Zimbabwean BioSafety Board for the field-testing of genetically modified materials; Bt maize and Bt cotton. The Bt would be tested on local varieties as well as on imported lines. It was only in 2001 that Bt maize trials were sanctioned. In 2002 a major policy question arose between the government and the World Food Programme on the importation of food aid. The only maize that the US government would donate was genetically modified. The BioSafety Board was given powers to advise and to supervise the regulation of the importation.

The regulation and control of biotechnology issues in Zimbabwe is the responsibility of the Bio Safety Board, established in 1998 to promote and regulate biotechnology in the country. This 12-member board uses a checklist of questions to assess a proposal and judge whether to approve or not. This board is assisted by a team of 30 trained inspectors, who are invited from different organizations to carry out checks on the crop trial process at several different stages in the growing season and oversee the final disposal of the trial crop. These inspectors have also been engaged by the government to oversee the containment and milling of imported GM maize in order to prevent its use.

Since the establishment of the Board several landmark events have occurred concerning GMO materials. In 1996 an illegally established Bt cotton trial was destroyed after it was discovered. In 2001 the small-scale field trials of Bt cotton and maize were approved. These trials were held at research stations in Kadoma and Mt Hampded, but failed to give meaningful results due mainly to management failures.

As part of the regulation process, various safety measures must taken around trial sites to ensure that the sites are controlled environments. Isolation distances of 400 m for maize and 100 m for cotton are stipulated. The trials are placed at research stations rather than conducted as on-farm trials in order to allow stricter monitoring. The placement of mesh wire around sites is meant as a deterrent to theft. In the case of maize the detasselling at flowering is meant to control the spread of pollen to other wild relatives. There is also a condition that all harvested crops should be completely destroyed. The Bt maize trial has to be cancelled as some of these regulations were breached and the Bt cotton trial was also scrapped.

Food aid

The 2002/3 season saw the whole Southern Africa region engulfed in a food crisis. With a national consumption requirement of 1.64 million metric tonnes Zimbabwe managed to produce less than half a million tonnes. The United States ensured that the only humanitarian maize on offer was GM. The Zimbabwe government referred the matter to the BioSafety Board for an opinion, and the Board engaged a consultative process with experts drawn from the ranks of civil society. In the meantime the Zambians had rejected the GM food aid as 'unsafe' and this put even more pressure on the Zimbabwe discussion. Tony Hall, the US ambassador to the FAO, issued a statement accusing states that refused of 'crimes against humanity'. The UN issued the following statement in support of USAID/WFP:

> *With respect to GM maize, soy flour and other commodities containing GMOs, FAO and WHO are confident that the principal country of origin has applied its established national food safety risk assessment procedures. FAO and WHO have not undertaken any formal safety assessments of GM foods themselves. Donors to the WFP have fully certified that these foods are safe for human consumption.*

> *Based on national information from a variety of sources and current scientific knowledge, FAO, WHO and WFP hold the view that the consumption of foods containing GMOs now being provided as food aid in southern Africa is not likely to present human health risk. Therefore, these foods may be eaten. The Organizations confirm that to date they are not aware of scientifically documented cases in which the consumption of these foods has had negative human health effects.*

> *In the specific case of maize, processing techniques such as milling or heat treatment may be considered by governments to avoid inadvertent introduction of genetically modified seed. However, it is not UN policy that GM grain used for food, feed, or processing should necessarily require such treatments.*

> UN statement on the use of GM foods as food aid in Southern Africa, Rome, 27 August 2002

The different Southern Africa states had different positions on the importation of GM materials. Of the six countries experiencing food shortages, only Swaziland had no objection to GM food. Lesotho, Mozambique, and Zimbabwe had insisted on GM maize being milled before distribution. Zambia barred the importation of all GM food until a team of its scientists touring overseas capitals completed a review on the safety of GM food. Malawi insisted that GM maize be milled during the planting season (IRIN News, 30 September 2002).

Obviously, for Zimbabwe, the rejection of the genetically modified maize was more than just a technical decision but rather became a highly politicized, nationalistic stance by the Zimbabwe government against the west. With the worsening food situation at hand, the milling option became unavoidable. However, many still saw the pressure by the developed countries not as sympathy for the starving people but rather as 'an opportunity to force GMO material down the throats of Africa'.

Based on the evidence at hand, the BioSafety Board decided there were no significant risks posed by consuming Bt maize. The main fear was that farmers would retain some of the grain for planting and as such a recommendation was made that all maize imports be milled immediately upon entry into Zimbabwe.

Why refuse GMOs?

A number of issues prevalent in the international GMO debate were raised by jurors participating in Izwi neTarisiro.

GM crops cannot co-exist with non-GMO crops without contaminating them. It is a known fact that pollen grains can be spread a long distance through open pollination by wind and insects. The potential exists for smallholder farmers to lose the qualities of their indigenous seeds through this contamination. This would be a great loss since African farmers have endeavoured to save the seeds they 'trust and know' over centuries. Another negative scenario would be that farmers found with GM crops growing on their farms might be prosecuted for violating patents. Patent owners might start demanding royalties or taking legal action.

There is concern that patents and copyright are given (mainly to companies in developed countries) too freely and on terms that unfairly penalize consumers, researchers, and small producers. Martin Khor, Director of the Third World Network, has spoken out against the present practice of patenting of genes, arguing that it is an abuse of the patent system since gene sequences are discoveries and not inventions (www.peoplesearthdecade.org, 4 October 2004).

Through the introduction of GM seeds, the small-scale farmers fear that they would ultimately lose their right to save or share seeds. Most GM seed manufacturing companies prohibit farmers from saving their on-farm produced seeds for coming sessions or sharing them with other farmers. This is done through elaborate contracts, agreements, and conditions which are imposed by GM seed producing multinational companies.

Over 80 per cent of the small-scale farmers in Africa today save their on-farm produced seeds for the next season. The farmers do this because they do not have enough income to buy seeds from companies for every planting season. Also, seed sharing is a cultural norm in many African communities.

African farmers have reacted with horror to the development of 'Terminator Technology' and 'Traitor Technology', two examples of Genetic Use Restriction Technologies (GURTs). 'Terminator Technology' is developed so that the planted GM seeds produce corresponding crops that produce sterile seeds, i.e. seeds which cannot germinate in the next season or at any other time. In 'Traitor Technology' the planted GM seeds produce corresponding crops that require application of certain chemicals to trigger (switch-on or turn-off) certain growth traits (e.g. germination, flowering, fruit ripening, leaf production, etc).

Any development or spread of the GURT technologies over the next 20 years could have significant negative impacts on the livelihoods of small-scale farmers in Africa. The farmers and country would wholly depend on companies for their seed supply. They would be forced to buy the 'switch-on' or 'switch-off' chemical stimulants for their crops. Jurors a raised the issue of litigation for possible harm from GMOs. The fact that the harmful effects may take long to manifest themselves further complicates this discussion. If it is the environment that has been contaminated, who would be liable – the planter, the breeder? And to whom would they be liable? Even during the jury discussions these questions were not resolved but it was felt that corporation should be regulated by a compensation framework should such damage occur. By virtue of the consumption pattern of maize in staple diets, concerns about food tastes were a key issue. In Zimbabwe maize-based food can be eaten up to three times a day and in large quantities. The main difference in consumption studies that had been done in the US was that corn in the United States was consumed more as a snack, which has gone through various processing phases. Still other jurors raised the issue of tampering with nature. They feared that 'such interferences with nature, as in genetically modified organism, may have been the one that resulted in such devastating problems like HIV and AIDS'.

Concerns about future effect on the health of the consumer were raised. A juror gave an example of 'when such chemicals as DDT and Garmatoxe were released they were hailed as wonder chemicals and they helped many at that time. It however took the world years before such chemicals were declared hazardous and banned. How safe will genetically modified foods be in the long term?'. The threat to human health that is posed by GMO materials is little known. Many fear that continuous uptake could lead to emergence of new viral strains or allegenicity that may be hard to control. This could lead to more toxins in the body, more allergic reactions, or the emergence of the new viral strains which may hard to control or manage. Emergence of new diseases has become a real threat in Africa where most people are poor and the medical services poor and fragmented. A living example is the experience with HIV/AIDS.

For Bt maize there is the promise of better stem bollworm resistance and so better yield and income to the farmer. Historically in Zimbabwe and indeed other semi-arid environments of Africa, this has never been a huge problem for smallholder farmers. Because of the dry environment, low rainfall, and poor nutrient availability bollworms rarely pose a significant threat in smallholder agriculture. The main threat for smallholder agriculture is low rainfall and a short season and drought. This was echoed by one farmer: 'You will all appreciate that a farmer cannot do without seed. So what are we going to do as we wait anxiously and indefinitely for the release of a drought tolerant maize seed variety?'

The GMO discussions at *Izwi neTarisiro* revealed the large extent of uncertainty both on the part of the specialist witnesses and the jurors. Because of the vulnerable position that developing countries find themselves in there is resignation that even when poor countries may have concerns these can be overridden by diplomatic and economic pressures from developed countries.

A key question, is who will make decisions on the range of concerns raised, and how will such debates be organized. These debates are live in Zimbabwe with organizations such as the Biotechnology Trust of Zimbabwe organizing public debates and consultations such as the *Izwi neTarisiro*. However, the mandate of the Biosafety Board remains narrow. In the final analysis their competence to address these other questions is doubtful, as they themselves admit. Yet, at the moment, there is no other route to address the wider set of concerns beyond the narrow technical issues of biosafety (Keeley and Scoones, 2003). There is no formal policy process looking at wider issues of agricultural change and strategy that are connected to biosafety regulation, beyond ongoing informal debate and lobbying by the range of actors from NGOs to industry.

Defending local knowledge, a local seed fair in Chivi district   *Source*: ITDG/Margaret Waller

5.2 Jury on International Property Rights (IPRs)

The discussion on IPRs was mainly an educational exchange for the jury. This discussion coupled with the jurors' own experience resulted in more insights on how seed and other IPR issues have come to be. The discussion centered mainly around seed as this directly affects smallholder farmers. The maize seed sector in Zimbabwe is mainly in the hands of the private sector with five big seed houses dominating the hybrid market. Research in maize is regulated and facilitated through the State Research and Specialist Services. Up until 2001 the production and release of Open Pollinated Varieties was restricted. Presently two varieties are available. The collapse of the commercial farming sector, which traditionally multiplied seed, saw an initial decline in seed availability and also the entry of smallholder farmers into seed production.

NGOs, with some support from the Ministry of Agriculture, are encouraging smallholder farmers to move into research and production of new seed varieties. Initiatives taken include the establishment of community seed banks and the introduction of smallholder farmers into seed production, both hybrid and open pollinated varieties. Farmers have also been advised not to underestimate the importance of their traditional seed. They have also been encouraged to enter into agreements with international research companies to share the rewards from the sale of improved seed varieties produced using knowledge originally obtained from their sector.

The problem of bio-piracy is considered real and there is documented evidence of such cases in the country. The problem is exacerbated by a culture that promotes free sharing of vital information, which leaves it open to be privatized by bio-prospectors at a later stage.

Issues of concern to smallholder farmers emanated from fears that global seed regulations and agreements, once applied in Zimbabwe, would prevent smallholder farmers from exchanging seed among themselves. As such there was a strong desire for governments push for local guidelines and regulations to protect local seed knowledge and prevent its privatization.

The issue of bio-piracy was found to be particularly disturbing in smallholder agriculture. Information and materials are collected and never acknowledged by outsiders and these are used for gain that never reaches the information source. To this the specialist witness talked of *'documents on the drawing board and intense debate involving government and stakeholders interested in the subject. This is also supported by international conventions on international property rights that are demanding the establishment of appropriate rules and regulations by governments, which protect farming communities from bio-piracy. These laws would firmly establish protocols to be followed by outsiders leading to the signing of agreements that are mutually beneficial to the parties concerned regarding the use of information obtained from farmers and sharing of the benefits.'*

Smallholder farmers expressed the hope that scientific advances would lead to better farming prospects, as they demanded that they had 'heard so much about research being conducted on a drought tolerant maize species. When is this coming to a conclusion and when is it going to be ready for use?'. The release of the variety would once again raise doubts on the benefit sharing regime proposed under the UN Convention on Biological Diversity where: *'the complexity of international property rights is evident. You will be interested to know that the seed you are referring to originated from Mexico. This was crossed with local strains and further researched on by a Kenyan research institute. The funds needed were provided by the Dutch. Now you can see for yourselves the daunting task we face, i.e. how do we determine fair sharing of benefits and who will be given the privilege to lead future research? This dilemma is a function of selfish tendencies by some of the stakeholders leading to prolonged debates and delays at the expense of expectant farmers.'*

Smallholder farmers claimed a voice and the right to conduct their own research as they consider it to be grossly unfair to consider that they have lost property rights on seed varieties, some of which originally belonged to them. They are left with the worry that by 2020 they will be completely restricted from doing anything by patent laws that are to a large extent granted and held by the corporations in developed countries.

The research on and the production of new seed varieties is not easy and requires considerable expertise. In most cases developing country governments do not have the scientists, funds and laboratory equipment needed. Zimbabweans have the farming experience and seed genetic material. This is why there is need for farmers to be given the option to produce own seed for their own consumption and not for commercial sale. This gives them the leeway to continue with their farming activities whilst the government is pursuing the issue of appropriate patents. Obviously there is a huge knowledge and information gap between developed and developing countries that is still to be bridged.

6
Conclusions on the Zimbabwean process

Most witnesses gave positive feedback on the *Izwi neTarisiro*, as a direct way of exposing policy makers to the voices of ordinary people. They were also struck by the extraordinary capacities of the 'ordinary' farmers. To quote Julius Magwagwa of BTZ, 'we often just generalize that these farmers don't know much but when given room and a conducive environment they even take a government minister head on. They felt that the process should be adopted and carried out more frequently on particular issues within government departments'.

The farmers themselves felt that the process was useful in establishing the principle that officials can be brought to answer for the weakness of implementation of policies on the ground. However some frustration was eventually voiced that since the hearings did not correspond to any single real policy document or process, the potential for follow up actions from this particular event might be limited. They also expressed some surprise that government officials and academics did not adequately respond to their brief of presenting a vision of agriculture in 2020. They laid out their existing plans and only turned to the vision element very briefly.

Here are some further comments by individual farmers and witnesses on the usefulness of the process:

- *It will be useful in engaging agro-dealers when they introduce new products to us.*
- *We can use it to inject new knowledge and assess it transparently with others.*
- *We can use it to deal with adjudication issues especially on land use at district level.*
- *Unlike our chaired meetings the process is more flexible and allows for in-depth questioning for thorough understanding.*
- *It allows faster dissemination of new ideas in a community.*
- *It helps to minimize conflict.*
- *It helps to create perfect understanding of a system and how it works and helps us build realistic expectations of ourselves.*

Reflections of the day's proceedings were conducted at the end of each day with the Oversight Panel members. These reflections were done to guide the process. On the last day Oversight Panel members did a formal evaluation of the process.

The quotes below reflect their view of the process stages:

1. On the ability of smallholder farmers (jury members) to discuss complex issues:

   Being largely a socio-economically disadvantaged group, the jurors' ability to internalize sometimes seemingly complex issues and ask pertinent questions was surprising. Their confidence especially in hotly engaging their farmer union leaders was further evidence of their ability to articulate issues affecting them.

   The emotive tones in reaction to their Union leaders were understandably due to long-existing communication problems. This reaction was however largely off-point as it did not relate to the process of policy formulation at hand.

A number of questions posed reflected that a donor-dependence is still very much alive. Farmers are still looking outward for solutions rather than for means to solutions they perceive to holistically address their lot. Farmers do not realise the power in them to effect change.

The level of engagement that resulted render this process a viable method of involving the marginalized at grass-roots, in consensus building for policy shaping.

2. On the role of specialist witnesses in reaching consensus:

It is very apparent that the witness component is essential in providing information that might otherwise remain unknown to the public. With information small-holder farmers are capable of asking pertinent questions to further clarify and make suggestions in view of their situation and concerns.

3. On the possibility of future application:

If acceptable, steps should be taken for this process to be mainstreamed as a tool in policy-making by Government and other organizations.

4. On the immediate use of the process:

Each of the participants having been so heavily involved over five days could utilise some of the skills, techniques, knowledge learnt in addressing development visions and strategies in their locality immediately upon returning.

5. On assessing the objectivity of the process:

The oversight panel needs to have an objective assessment tool on which to base their judgement of the process. This is important if the method is to be repeated by others, for comparison.

Throughout the week of hearings in Harare, witnesses presented a model of the policy formulation process under parliamentary democracy – how prices are fixed, how legislation is drafted, etc. The citizens' jury method is intended to open up the scrutiny of policy issues outside closed parliamentary committees of experts. However, perhaps this is not a true reflection of the policy formulation process in twenty-first century Zimbabwe. One would not go to a parliamentary committee in Zimbabwe today to observe the routine development of policy options for the future of agriculture. Policy specialists who presented the formulation of agricultural policies to the jury may therefore have given false expectations of likely policy influencing options.

On the other hand, given their exposure to unrealistic claims about government policies that bear no relation to grim local realities, farmers left the week's proceedings and returned to their districts with a clearer picture of the deterioration of government services and armed with a range of new ideas on livelihood options.

7
Ways forward

The successful adaptation of the Farmers' Jury approach to the Zimbabwean context provides a good starting point for developing a strong and prosperous smallholder agricultural sector in Zimbabwe. Not only did the process highlight the importance of citizen participation but it also demonstrated that open dialogue and consultation can improve the quality of policies and programmes meant to benefit smallholder farmers. The dialogue that ensued between farmers and high-ranking government officials revealed some of the major shortcomings in previous and existing smallholder agricultural policies. It would be useful therefore to consider the way forward at two levels: (1) the future of the jury process itself in the Zimbabwean context; and (2) the specific issues and challenges raised in the jury.

While many people have emphasized the need for participatory policy formulation in Zimbabwe, there have been very scanty efforts at testing methods for improving citizen participation in policy formulation. The jury approach provides a unique window for a practical model that can be adapted to a number of sectors in Zimbabwe to bridge the disconnection between farmers' needs and policy priorities. The partnership approach that was used to test the concept proved very successful and future efforts to expand or replicate the process would need to adopt similar principles for credibility and wider buy-in. The facilitation of this process by an NGO was very useful in terms of accessing international support and guidance to inform the process on the ground. A comparative analysis of different experiences needs to be undertaken to build a sufficiently broad knowledge base to enable other mainstream actors to take the lead in the expansion of the jury approach to participatory policy formulation.

In a context where institutional capacities are freezing due to financial and administrative bottlenecks, a more targeted capacity building process is needed. The jury could even be applied at a micro-institutional level to resolve a problem being experienced. It is always important to create learning opportunities for a wide range of actors throughout the planning and implementation of a jury process, to enable follow up on issues by all participating stakeholders.

If wider uptake of the jury process is intended, the jury process must not only be seen as a tool for enabling compatibility of policies with people's needs but as a process for institutional renewal and growth. Interestingly, the jury process brought out some useful avenues for institutional development and growth of key institutions such as ZFU and AREX. If well implemented, the jury has the potential to consolidate the legitimacy and credibility of institutions that serve smallholder farmers. Juries can also improve the focus of institutions that work with farmers in Zimbabwe. These benefits should provide strong incentives for the uptake of the jury approach for institutions working with smallholder farmers in Zimbabwe.

It would be much more useful if the idea of a jury were mooted through mainstream government agencies although it is not best practice for the initiating agency to facilitate its own process. The principle would be to find a neutral facilitator able to command the respect of both farmers and service organizations. More practical examples however, would accelerate uptake. Unfortunately, sponsorship for practical models of participatory policy formulation is very limited due to the flight of donors from Zimbabwe to neighbouring countries. Hence, opportunities for responding to specific demands for the design and implementation of farmer juries need to be identified.

At the farmer level, a fundamental challenge of limited institutional capacity in farmer representation also emerged. Farmer representative organizations can also use the jury approach to prioritize issues and validate their focus. It is a tool that can strengthen the policy content and directions of these organizations. The output of a jury in terms of advocating for alternative policy positions would be a strong input into the policy formulation process which is normally based on strength of position papers from various interest groups. The output of jury processes is usually very emphatic and clear in terms of positions being supported or rejected by farmers.

NGOs have usually been the main drivers of participatory initiatives. It is important they take this role in its most facilitating and enabling form while building the capacity of local organizations to play a leading role. NGOs must channel their resources to strengthening local institutional competences to drive processes of change that have a bearing on their lives. Hence regular planning and reflection meetings should be regularly convened to clearly define the limits of the NGO facilitation role.

The debate on the land reform issue in Zimbabwe highlighted a number of challenges. Farmers' interests and priorities are not at the core of the land reform programme as farmers still raised questions on the content and overall direction of the programme. The isolation of farmers from the programme has been further worsened by the defective planning and implementation strategies which have even ignored the role of local traditional leaders. In many cases, the role of traditional leaders was seriously undermined. Many of the clarification questions raised by farmers during the jury process would have been addressed at planning stage if a participatory policy formulation approach had been adopted. The land reform process also seems to be disconnected and not well integrated with other sectoral initiatives. The land reform process cannot be implemented successfully without forging strategic links with other key sectors. The debate managed to highlight areas for future dialogue between farmers and responsible organs of government implementing the land reform programme. More dialogue and communication as well as improved information flow would resolve some of the outstanding issues. The setting up of a public enquiry system linked to small scale farmers would go a long way in building dialogue between farmers and implementing and support organizations.

A major issue of concern however, revolves around the capacity of farmers to raise issues that affect them. There seems to be lack of confidence by farmers to view themselves as agents for change in their own right. Farmers need to be empowered to be more proactive in raising issues that affect them. It would be useful for external support to target strengthening farmer decision making support systems for them to come up with informed decisions and advocacy strategies.

The HIV/AIDS issue is a very emotive subject for smallholder farmers. They are the hardest hit by the pandemic and yet many centrally driven initiatives have failed to reach them. The jury debates illustrated the confusion that still characterize national responses and the crisis of expectations often created through publicity of some initiatives that remain hanging above the suffering communities. It will be useful to undertake a more systematic review of HIV/AIDS initiatives. The jury approach could be adapted to validate the national vision of the HIV/AIDS programme in Zimbabwe. A multi-sectoral coordinated response is needed at community level to avoid the creation of confusion and conflict of interventions.

Peasant farmers are still left out of the GMO debate and their natural response to hearing about the issue is one of fear and nervousness. This is understandable given the lack of public information on GMOs. There is a need to improve information sharing with farmers on GMOs so that they can participate meaningfully in the debate. Equally, although Intellectual Property Rights are topical globally, knowledge and information gaps have excluded smallholder farmers from this debate.

Overall, the Farmers' Jury process in Zimbabwe demonstrated that there is some space to promote open deliberative processes. Building shared commitment to participatory policy formulation is achievable through open dialogue and collaborative planning. Understanding the context (institutional, social, economic, and political) is critical to the application of the farmer jury methodology in Zimbabwe.

Appendices

Appendix 1   The verdict

We, having heard evidence and deliberated from 24 to 28 February deliver the following verdict:

1. **Water and agriculture**
   We desire:
   - development and expansion of communal irrigation scheme facilities in rural areas;
   - awareness, information, and promotion of soil and water conservation for smallholder farmers;
   - equal access to such irrigation technologies as drip irrigation;
   - the full participation of all users in the review of irrigation water costs and related legislation;
   - the construction of dams, where potential exists, to augment rains;
   - tree planting and the promotion of such techniques that raise the water table;
   - full participation and involvement of smallholder farmers in the allocation of water permits in catchment councils.

2. **Rural livelihood options**
   We desire:
   - the promotion of processing infrastructure for natural resources e.g. mazhanje jam, masau;
   - local level value addition e.g. peanut butter processing;
   - the creation of incentives that can lure investment in smallholder agriculture;
   - the strengthening of rural urban linkages.

3. **Research and extension**
   We desire:
   - the promotion of extension systems that encourage group approaches;
   - sufficient knowledge of seed breeding for both modern and traditional varieties;
   - the presence of committed and dedicated extension workers for backstopping support;
   - farmer-led farm site research driven by the farmer's research agenda;
   - extension approaches that value the farmer's knowledge;
   - the promotion of farmer to farmer extension approaches that uses farmer innovators as community level extensionists;
   - full cycle farmer training using such training as farmer field schools.

4. Farming systems

We desire:
- optimum yields from the use of farm yard manure and such soil fertility systems that are locally available;
- the formation of strong farmer groups;
- the promotion of the conservation of traditional crops and livestock;
- value addition of farm produce;
- crop rotation;
- mixed and intercropping;
- the promotion of growing a wide range of crops and varieties;
- the promotion of the use of local plants as repellents;
- research into the use of such chemical repellents to improve them;
- the availability of adequate draft animals;
- local capacity to make tools and equipment;
- the promotion of systems that do not harm the local ecosystems.

5. Education and training

We desire:
- documentation and dissemination by farmers of where we came from, our plans, and our expectations;
- farmers able to read and write;
- the farmer should choose what they want to learn and decide on what topics to specialize in;
- the linkage of fellow farmers to fellow farmers, researchers, academic institutions and such sources of information and knowledge;
- the promotion of farmer exchange and exposure to increase knowledge;
- self reliance;
- the availability and use of information and communication technologies by farmers.

6. HIV/AIDS and labour

We support:
- home-based care programmes for people affected by HIV;
- the provision of funds specifically for establishing short-term and long-term project targeting groups or families of people living with HIV/AIDS;
- funds for the support of those affected by AIDS should go directly to the people affected;
- inclusion of AIDS education in the current education syllabus from primary school level;
- the training and promotion of community level educators;
- promotion of good values, e.g. through voluntary virginity tests. [This was advocated by elder members of the jury who gave experiences of how moral values were valued and upheld in their day. There was no unanimity in this as others felt it was unfair for girls to be targeted for the virginity tests];
- the full support with training and materials of caregivers.

We oppose:
- heavy promotion of condoms at the expense of moral values.

7. Intellectual property rights

We desire:
- laws that protect our seed from being used to develop hybrids and protect the exploitation of our natural resources for corporate gain;

- developing countries should be consulted when laws are being formulated;
- smallholder farmers should be allowed to trade their own seed;
- smallholder farmers should be allowed to produce their own seed;
- there should be laws that allow for compensation and consent where our knowledge is to be used.

We oppose:
- the surrender of seed production rights to large corporations.

8. Natural resource management
We support:
- the strengthening of by-laws that promote the conservation of natural resources;
- the local level issuance of licenses for gold panning;
- the strengthening of laws that promote the sustainable use of vulnerable environments such as stream banks and sacred places;
- the consolidation of all laws that have to do with environmental management;
- the compensation of people affected by local developments and these displaced should benefit from the development.

9. Land
We support:
- the allocation of land to farmers with a proven track record of farming.

We oppose:
- the destruction of natural resources within resettled lands.

10. Genetically modified organisms
We support:
- full awareness and education about genetically modified organisms;
- more research into the pros and cons of genetically modified organisms;
- there should be compensation where people are affected by genetically modified organisms.

We oppose:
- the use of genetically modified organisms for food as there is no guarantee about their safety and effect in future;
- the use of genetically modified organisms as they may have an effect on the environment.

11. Credit
We support:
- farmers should have their own community banks;
- the availability of water and appropriate technology which allows farmers to be self reliant from loans.

We oppose:
- the use of livestock as collateral.

12. Vision of markets by 2020
We support:
- access to local markets and other markets;
- availability of market information;

- full representation in contract and price negotiations.

We oppose:
- dictation of unfair prices to farmers.

13. **Vision of farmer institutions by 2020**
We desire:
- farmers be linked to retail and marketing institutions;
- elected union democratic leadership;
- farmer commodity association interest groups.

We oppose:
- voices that speak for farmers without consultation.

Appendix 2 Summary of jurors

	Province	District	Name of Farmer	Age (years)	Family size	Sex	Dryland	Irrigated	Livestock Cattle	Sheep	Goats	Poultry
1	Mashonaland Central	Guruve	Rosemary Tichareva	42	6 children	F	12 acres		4	0	3	15
2	Mashonaland Central	Muzarabani	Felix Ngorovhani	50	4 children	M	15 acres		1	0	3	15
3	Mashonaland Central	Mutoko	Danai Agrippa Nyakanyanga	28	2 children	M	32 acres	5.5 acres	8	0	25	25
4	Mashonaland East	Rushinga	Jones C Munemo	51	6 children	M	10 acres		6	12	6	10
5	Mashonaland Central	Buhera	Richman Newende	37	6 children	M	4.5 acres	1 acre	2	0	7	8
6	Mashonaland East	Hwedza	Jane Mashonganyika	49	9 children	F	8 acres	1 acre	6	0	12	15
7	Manicaland	Chimanimani	Jabulani Marangwana	54	5 children	M	1.5 hectares	0.4 hectares	0	0	0	0
8	Masvingo	Chivi	Zvizvai M Gunge	62	6 children	M	4 hectares		14	0	22	20
9	Matebeleland North	Tsholotsho	Thandekile Moyo	36	4 children	F	10 acres		3	0	3	25
10	Mashonaland West	Kariba	Muchena Chipangura	32		F	4 acres		0	0		
11	Midlands	Zvishavane	Boas Mawara Munyani	56	12 children	M	16 hectares		26	0	0	95
12	Mashonaland Central	Mt Darwin	Rosemary Mukwena	43	5 children	F	10 acres		15	0	3	25
13	Manicaland	Buhera	Chipo Murimwa	32	1 child	F	8 acres		5	0		7
14	Manicaland	Mutasa	Elia 40 Matsikira	40	18 children	M	9.9 hectares		2	0	29	0
15	Manicaland	Nyanga	Michael Nyamutowa	39	4 children	M	2 acres	2 acres	4	0	6	50
16	Mashonaland East	Mudzi	Mike Nyamukondiwa	46	5 children	M	4 acres		4	0	3	

Appendix 3   Summary of expert witnesses

Summary of Expert Witnesses Invited to the Workshop

A vision for HIV/AIDS in smallholder agriculture	Dr Kadenge/ Mr Mupambirei Kufakunesu	University of Zimbabwe National Aids Council
A vision of rural livelihoods	Dr Mtetwa	Director Enda Zimbabwe
A vision on water and agriculture	Dr Nyagumbo Chitsiko	University of Zimbabwe
A vision for farming systems	Dr Mpofu	Director Department of Engineering
A vision of agricultural research and extension	Mr OJ Zishiri Mr Chiteka	Director AREX University of Zimbabwe
A vision of training and education	Mr HBK Hakutangwi	Swedish Cooperative Centre
A vision of intellectual property rights	Mr A Mushita	Director Commutech
A vision on common property and NRM	Ms Chasi Ms B Jiji	Director Dept of Natural Resources Safire
A vision on smallholder farming and GMOs	Mugwagwa na Masimbe Mr Mafa	Biotechnology Trust of Zimbabwe Science and Technology Department
A vision of farmer institutions by 2020	Mr Mautsa Mr Tsikisai	Director ICFU Director Zimbabwe Farmers Union
A vision on Access to credit by 2020	Mr A Jaure	SCC/ZFU

Bibliography

Chaumba, J., Scoones, I. and Wolmer, W. (2003) New politics, New livelihoods: changes in the Zimbabwean lowveld since the farm occupations of 2000, *Sustainable Livelihoods in Southern Africa Research Paper 3*, Institute of Development Studies.

Coulter, J., Goodland, A., Tallontire, A., and Stringfellow, R. (1999) Marrying farmer cooperation and contract farming for service provision in a liberalising Sub-Saharan Africa, *Natural Resource Perspectives*, 48. Overseas Development Institute: London.

Derman, B. and Gonese, F. (2003) Water Reform in Zimbabwe: Its Multiple Interfaces with the Land Reform and Resettlement, in: Roth M. and Gonese, F., *Delivering Land and Securing Rural Livelihoods: Post Independence Land Reform and Resettlement in Zimbabwe*, CASS/University of Wisconsin-Madison.

FAO (2001) The impact of HIV/AIDS on food security. Twenty-seventh session of the Committee on Food Security, Rome 28 May–1 June 2001. Available at http://www.fao.org/docrep/meeting/003/Y0310E.htm. Accessed September 2004.

Gonese, F. and Mukora, C.M. (2003) Beneficiary Selection, Infrastructure Provision and Beneficiary Support, in: Roth M. and Gonese, F., *Delivering Land and Securing Rural Livelihoods: Post Independence Land Reform and Resettlement in Zimbabwe*, CASS/University of Wisconsin-Madison.

International Fund for Agricultural Development (IFAD) (2001) *The challenge of ending rural poverty*, Rural Poverty Report 2001.

Irwin A. and Wynne, B. (1996) *Misunderstanding Science*, Cambridge University Press, Cambridge.

Keeley J. and Scoones, I. (2003) Context for regulation of GMOs in Zimbabwe, IDS Working Paper 190.

Marongwe, N. (2003) The fast Track Resettlement and Urban Development Nexus, in Roth M. and F. Gonese, *Delivering Land and Securing Rural Livelihoods: Post Independence Land Reform and Resettlement in Zimbabwe*, CASS/University of Wisconsin-Madison.

Mundeiri, B.A. (2003) Land Administration and Decentralization Delivery of Extension Services to Land Reform Beneficiaries, in: Roth M. and Gonese, F., *Delivering Land and Securing Rural Livelihoods: Post Independence Land Reform and Resettlement in Zimbabwe*, CASS/University of Wisconsin-Madison.

Ncube, D. (2003) Role of Private Land Markets in Delivering Land and Beneficiary Support Service, in: Roth M. and Gonese, F., *Delivering Land and Securing Rural Livelihoods: Post Independence Land Reform and Resettlement in Zimbabwe*, CASS/University of Wisconsin-Madison.

Pimbert, M.P. and Wakeford, T. (2002) *Prajateerpu. A citizens jury/scenario workshop on food and farming futures for Andhra Pradesh, India*, IIED, London.

Pimbert, M.P. and Wakeford, T. (2003) An introduction to *Prajateerpu, a citizens jury/scenario workshop on food and farming futures for Andhra Pradesh, India*, in PLA Notes, 46.

Runge, C.F., Senauer, B., Pardey, P.G. and Rosengrant, M.W. (2003) *Ending hunger in our lifetime: food security and globalization*. International Food Policy Research Institute and The Johns Hopkins University Press.

SADC/FANR Vulnerability Assessment Committee (2003) *Towards identifying impacts of HIV/AIDS on food insecurity in Southern Africa and implications for response: findings from Malawi, Zambia and Zimbabwe*, Harare, Zimbabwe.

Stringfellow, R., Coulter, J., Lucey, T., McKone, C. and Hussain, A. (1997) Improving the access of smallholders to agricultural services in Sub-Saharan Africa: farmer cooperation and the role of the donor community. *Natural Resource Perspectives*, 20, Overseas Development Institute, London.

The Economist (2004) The cost of AIDS: an imprecise catastrophe, 20 May.

Topouzis, D. (2003) *Addressing the impact of HIV/AIDS on Ministries of Agriculture: focus on eastern and southern Africa*, A joint FAO/UNAIDS Publication, Rome.

UNAIDS (2002) http://www.unaids.org/. Accessed September 2004.

UNAIDS (2003) http://www.unaids.org/. Accessed September 2004.

UNAIDS (2004) http://www.unaids.org/. Accessed December 2004.

Wakeford, T. and Pimbert, M.P. (2004) Prajateerpu, Power and Knowledge: The politics of participatory action research in development (posted at http://www.prajateerpu.org).

Whiteside, M. (1998) Encouraging Sustainable Small Holder Agriculture in Zimbabwe, DFID, mimeo.

World Bank (1999) Rural Producer Organisations: their Contribution to Rural Capacity Building and Poverty Reduction, presented by the Rural Development Department, Washington, D.C., 28–30 June.

World Bank (2001) World Development Report 2000/2001: Attacking Poverty, The World Bank Washington D.C..

World Bank Group (2004) Millenium Development Goals. Available at http://www.developmentgoals.org/. Accessed December 2004.